Double-Star Astronomy: Containing the History of Double-Star Work: Computation of Orbits and Position of Orbit-Planes; Formulae in Connection with Mass, Parallax, Magnitude, Etc

Thomas Crompton Lewis

DOUBLE-STAR ASTRONOMY:

CONTAINING

THE HISTORY OF DOUBLE-STAR WORK;

COMPUTATION OF ORBITS AND POSITION OF
ORBIT-PLANES;

FORMULÆ IN CONNECTION WITH MASS,
PARALLAX, MAGNITUDE, Etc.

BY

T. LEWIS, Sec.R.A.S.,
ROYAL OBSERVATORY, GREENWICH.

LONDON:

PRINTED AND PUBLISHED BY TAYLOR AND FRANCIS,
RED LION COURT, FLEET STREET, E.C.

1908.

[*From* 'The Observatory,' 1908.]

DOUBLE-STAR ASTRONOMY.

In attempting the history of any subject there is always the same initial difficulty to confront—where to begin. There is an inherent tendency to push back the date as far as possible, and sometimes a little further. This is true in the present instance. Stars must for ages have received attention because of their apparent proximity to each other. I think, however, we may fairly claim the year 1779 as the true beginning of double-star astronomy. In this year Sir W. Herschel commenced definitely the work which finally led to the discovery of the real nature of these neighbouring suns. True it is that he had at this time another purpose in view, and it is very difficult, if not impossible, to say when the real meaning of his labours disclosed itself to him. We might claim another century, for Riccioli noted ζ Ursæ Majoris as double about the middle of the 17th century, while Huyghens saw θ Orionis quadruple in 1656 [*], and the duplicity of γ Arietis was detected by Hooke in 1664 [†].

In addition to these it may be of interest to give the date of discovery of a few others which have become well known in our time :—

	Discovered by	
α Crucis	Fontenay	1685 [‡]
α Centauri	Richaud	1689 [§]
γ Virginis	Bradley	1718
Castor	Bradley	1719
61 Cygni	Bradley	1753
ζ Cancri	Mayer	1756
ϵ Lyræ	Maskelyne	1765
α Herculis	Maskelyne	1777
70 Ophiuchi	Herschel	1779
ξ Ursæ Majoris	Herschel	1781

The discovery of α Centauri is usually ascribed to Feuillée [||] in 1709, from the fact that he was the first to give an estimate of

---

[*] Huyghens, 'Opera,' p. 540.   [†] Phil. Trans. No. 4, p. 108.
[‡] Histoire de l'Académie depuis 1686-1699, t. ii. (Paris, 1733) p. 19.
[§] Mém. de l'Acad. depuis 1686-1699, t. vii. 2 (Paris, 1729) p. 206.
[||] Journal des Observations physiques, t. i. (Paris, 1714) p. 425.

A

the relative positions of the components. The remarks of both these observers are interesting. Richaud says :—" Regardant à l'occasion de la Comète plusieurs fois les pieds du Centaure avec une lunette d'environ douze pieds, je remarquai que le pied le plus oriental et le plus brillant étoit une double étoile aussi bien que le pied de la croisade ; avec cette différence, que dans la croisade une étoile paraît avec la lunette notablement éloignée de l'autre ; au lieu qu'au pied du Centaure, les deux étoiles paraissent même avec la lunette presque se touche ; quoique cependant on les distingue aisement."

These stars were regarded not as in real proximity, but accidentally in the same visual line—one far beyond the other; and this belief accounts for the fact that such pairs were not searched for, their discovery being accidental. We find that immediately their true character was only even suspected, the number of known systems increased at a very rapid rate. It is true that Lambert in 1761 * maintained that the stars were suns analogous to our Sun, and were accompanied by a retinue of planets and comets; and the Rev. John Michell in 1767 † found by the application of the theory of chances that the grouping of stars was scarcely accidental, and that double stars found by the telescope were for the most part binary systems. These, however, were mere speculations of one or two individuals.

In 1779 appeared a small book entitled " De novis in Cœlo sidero Phænominis in miris stellarum fixarum comitibus," by Christian Mayer ‡, wherein he speculates as to the possibility of small suns revolving around larger ones, but of course he could give no evidence. Mayer was a Jesuit Father observing at Mannheim with an 8-foot Bird mural quadrant, using a power of 85. He collected all double stars known down to the year 1781, and by adding his own discoveries formed a list of 89 pairs. This Catalogue was published in Bode's ' Astronomische Jahrbuch,' 1784 : and as this may not be easy of access the following table (p. 90) will show the form employed (the stars are, of course, selected from various parts of the complete list).

It will be seen that the quantities are very rough and afford no basis for his speculations. Indeed his remarks show that he was rather thinking of proper motions, for he says :—" Double stars are those which are single to the naked eye, but which are separated by a less or greater magnification, often only by means of a very good telescope, into two stars a few seconds apart. . . . By means of careful observation of double stars it is possible to discover proper motions." The positions, in the last column, of the

* 'Cosmologische Briefe über Einrichtung des Weltbaumes,' J. H. Lambert, 1761.

† Phil. Trans. vol. lvii. pp. 231-264, " An Inquiry into the probable Parallax and Magnitude of the Fixed Stars from the Quantity of Light which they afford us. . . ."

‡ Christian Mayer, 1709-1783.

Extract from Mayer's List.
('Astronomische Jahrbuch,' 1784.)

	Grösse.	Gerade Aufst.	Abweichung.	Unterschied		Abstand.	Stellung des Kleinern.
				in der Aufst.	in der Abw.		
		G. M.	G. M.	Sec.	Sec.	Sec.	
γ Widder	beyde 5 ter	25 22	18 13 N.	3	12	12	S.W.
γ Andromedæ ...	2 und 6 ter	27 36	41 16 N.	14	6	12	N.O.
α Widder	2 und 9 ter	28 40	22 24 N.	o	2	2	S.
Castor	2 und 9 ter	110 7	32 22 N.	10	4	9	N.W.
ζ Cancri	7 und 8 ter	119 52	18 19 N.	o	8	8	S.
γ Virginis	beyde 5 ter	187 37	0 13 S.	7	6	9	S.O.
ε Leyer	6 und 8 ter	279 13	39 27 N.	3	3	4	N.O.
ς Leyer	beyde 6 ter	279 13	39 24 N.	o	2	2	S.

comes relative to the principal star are mere estimations. Measuring the angle with respect to the meridian was not possible nor considered important until the invention of Herschel's "revolving micrometer" in 1779.

Having disposed of these preliminaries, we can now turn our attention to Herschel's work; and for this purpose we must begin with his paper on the "Parallax of Fixed Stars," read before the Royal Society, 1781, December 6.

He points out that "to find the distance of fixed stars has been a problem which many eminent astronomers have attempted to solve"; and that the cause of non-success was due in a great measure to the inherent difficulties of the methods adopted. In general, the method of zenith-distances labours under the following difficulties: refractions imperfectly known; change of Earth's axis arising from nutation; precession and other causes not accurately known.

He then suggests the following method:—"Let O and E be two opposite points of the annual orbit, taken in the same plane with two stars, *a* and *b*, of unequal magnitudes. Let the angle *aOb* be observed when the Earth is at O, and let the angle *aEb* be also observed when the Earth is at E. From the differences in these angles, if any should be found, we may calculate the parallax of the stars, according to a theory that will be delivered hereafter. These two stars, for reasons that will soon appear, ought to be as near each other as possible, also to differ as much in magnitude as we can find them."

He then goes on to say that Galileo and others had suggested and tried this method, but had not avoided those errors because they had not chosen stars very close. The present method he shows to be free from the defects above enumerated, and states that parallaxes of o″·1 could be detected.

It is remarkable that this search after parallax should have been so prolific in discoveries, and yet of so little value to the main object. Bradley, labouring to obtain parallaxes, discovers the aberration of light and the nutation of the Earth's axis, and then Herschel, by the fact of observing two close stars in place of distant ones, lights upon the most unexpected and fruitful discovery of binary stars.

To return to Herschel's paper and the real ground-plan of the department of double-star work. The superstructure is not what the architect intended, and Herschel was himself the first to point out the discrepancy :—" As soon as I was fully satisfied that in the investigation of parallax the method of double stars would have many advantages above any other, it became necessary to look out for proper stars. This introduced a new series of observation. I resolved to examine every star in the heavens with the utmost attention and a very high power, that I might collect such materials for this research as would enable me to fix my observations upon those that would best answer my end. The subject has already proved so extensive and still promises so rich a harvest to those who are inclined to be diligent in the pursuit, that I cannot help inviting every lover of astronomy to join with me in observations that must inevitably lead to new discoveries. I took some pains to find out what double stars had been recorded by astronomers : but my situation permitted me not to consult extensive libraries, nor indeed was it very material ; for as I intended to view the heavens myself, Nature, that great volume, appeared to me to contain the best catalogue upon this occasion.. However, I remembered that the star in the head of Castor, that in the breast of the Virgin, and the first star in Aries had been mentioned by Cassini as double stars. I also found that the nebula in Orion was marked in Huyghen's 'Systema Saturnium' as containing several stars, three of which (now known to be four) are very near together.

" With this small stock I begun, and in the course of a few years' observations have collected the stars contained in my catalogue."

Some further remarks are interesting as showing the non-diffusion of useful knowledge at this period, and how extremely meagre and unsatisfactory were the few statements extant. Herschel's own notes, under exactly similar circumstances, I shall give presently. He says that " When at the Royal Observatory in 1781 the Astronomer Royal* showed me α Herculis as a double, stating that he had discovered it some years since. . . . . Mr. Hornby at Oxford mentions π Boötis as a double. . . . . It is a little hard on young astronomers to be obliged to discover over again what has already been discovered " †.

* Maskelyne.   † W. Herschel, 1738-1822.

In this particular instance, however, there are few who could bring themselves to sympathise with Herschel. Had he even known of Mayer's catalogue, he might have been content to use these stars, and then we should have lost all his discoveries, all his measures, and with them also the early discovery of binary stars. This paper literally teems with the most useful information, and I cannot resist quoting still further from his notes on observing, which most double-star observers even now either have adopted or have fallen into the habit of using. And we must bear in mind that Herschel had no clock-work to keep his stars in the field; this had to be done with his hands.

"In settling the distances of double stars, I have occasionally used two different ways. Those that are extremely near each other may be estimated by the eye, in measures of their own diameters. For this purpose their distance should not much exceed two diameters of the largest, as the eye cannot so well make a good estimate when the interval between them is greater. This method has often the preference to that of the micrometer: for instance, when the diameter of a small star, perhaps not equal to half a second, is double the vacancy between the two stars. Here a micrometer ought to measure tenths of a second at least, otherwise we could not, with any degree of confidence, rely on its measures."

The remaining part of the paper is very interesting, but is more distinctly concerned with parallax.

In 1782, Jan. 10, Herschel published a catalogue of double stars for the use of observers studying parallax. These he divides into six classes :—

1. Stars requiring good telescopes and favourable circumstances to separate.
2. Stars proper for estimation by the eye or very delicate micrometer measures.
3. All stars more than 5″ and less than 15″ apart, and cannot be looked upon as free from the effects of refraction, &c.

"The 4th, 5th, and 6th classes contain double stars that are 15″ to 30″, 30″ to 60″, and 60″ to 120″ or more apart. Though these will hardly be of any service for the purpose of parallax, I have thought it not amiss to give an account of such as I have observed; they may, perhaps, answer another very important end . . . . Several stars of the first magnitude have already been observed, and others suspected, to have a proper motion of their own; hence we may surmise that our Sun, with all its planets and comets, may also have a motion towards some particular part of the heavens."

Here, again, observations of double stars may serve for a purpose other than parallax; but no idea as to their real future use seems to have been entertained. The determination of proper motions from micrometric measures is not regarded now with so much.

favour, and yet I believe these observations capable of yielding very accurate results.

To return, this first catalogue of 269 double stars is of great importance, not only for the observations themselves, but for their exhaustive character and orderly arrangement. As an example of the amount of information and the thoroughness which characterized Herschel's work, one excerpt from the catalogue will suffice :—

"α Geminorum. Fl. 66. April 8, 1778.

"Double. A little unequal. Both W. The vacancy between the two stars, with a power of 146, is 1 diameter of S; with 222, a little more than 1 diameter of L; with 227, $1\frac{1}{2}$ diameters of S; with 460, near 2 diameters of L (see fig. 6); with 754, 2 diameters of L; 3168 the interval extremely large, and still pretty distinct. Distance by micrometer 5"·156. Position 32° 47' n preceding. These are all a mean of the last two years' observations, except the first with 146."

The first class contained 17 stars, the second class 38, and amongst them the following :—

First Class.		Second Class.	
Star.	First observation.	Star.	First observation.
ε Boötis	1779, Sept. 9	α Geminor.	1778, April 8
ξ Ursæ Maj.......	1780, May 2	α Herculis	1779, Aug. 29
σ Coronæ	1780, Aug. 7	ρ Herculis	1779, Aug. 29
η Coronæ	1781, Sept. 9	ε Lyræ	1779, Aug. 29
ζ Cancri	1781, Nov. 21	ζ Aquarii	1779, Sept. 12

So far as I can ascertain, the first star measured belonged to the third class, viz. θ Orionis, 1776, November 11. After this paper had been read, Herschel saw Mayer's memoir, " De novis in Cœlo sidero Phænomenis," which is contained in the 4th volume of the ' Acta Academiæ Theodoro Palatinæ,' and acknowledges in a postscript that Mayer had the ideas before him. He also says that he purposely used the expression " double star " in preference to such words as comes, companion, satellite, " because, in his opinion, it is much too early to form a theory of small stars revolving around larger ones." In the words of W. Struve, these speculations " n'obtinerent point, à cette époque, l'approbation de l'astronome calme de Slough." He found that 31 of Mayer's stars were not in his catalogue.

An important paper of this calibre naturally set astronomers to work, and in the ' Philosophical Transactions,' 1783, November 27, we find a paper by Mr. Michell, "On the Means of Discovering the Distance, Magnitude, &c. of the Fixed Stars in Consequence of the Diminution of the Velocity of Light," in which he says:— "The very great number of stars that have been discovered to be double, triple, &c., by Mr. Herschel, if we apply the doctrine of chances, cannot leave a doubt with anyone, that by far the greatest part, if not all of them, are systems of stars so near to each other as probably to be liable to be affected sensibly by their mutual gravitation; and it is therefore not unlikely that the periods of the revolutions of some of these about their principals (the smaller ones being, upon this hypothesis, considered as satellites to the others) may some time or other be discovered." He then proceeds to show that by comparing the brightness of Sirius with that of the Sun, through the intermediary of Saturn at a time when the ring was invisible and so almost identical with Sirius, the parallax of the latter is somewhat under $1''$. He gives the probability of the closeness of α^1 and α^2 Capricorni ($3'$ apart) as due to accident as $\frac{1}{80}$, and the grouping of the Pleiades as $\frac{1}{500,000}$. This, so far as I gather, is the first intimation of the coming discovery.

Herschel meanwhile had been hard at work preparing a second catalogue of 434 double stars, which he presented to the Royal Society on December 9, 1784, remarking that "The present collection is much more perfect than the former; almost every double star in it having the distance and position of its two stars measured by proper micrometers." The paper is full of careful directions for finding the stars and remarks on observing generally. This catalogue brings the numbers in the first and second classes to 97 and 102 respectively.

In a later paper "On the Construction of the Universe," Herschel does not mention that he entertained any suspicion of the binary character of double stars, and the first glimpse we get of such an idea is in a paper read before the Royal Society, 1802, July 1, in which he places all celestial objects under 12 heads. In the first class he has all insulated or single stars; in the second, double or binary stars :—"The union of two stars, that are formed together in one system, by the laws of attraction.... If a certain star should be situated at any, perhaps immense, distance behind another, and but very little deviating from the line in which we see the first, we should then have the appearance of a double star. But these stars being totally unconnected would not form a binary system. If, on the contrary, two stars should really be situated very near each other, and at the same time so far insulated as not to be materially affected by neighbouring stars, they will then compose a separate system, and remain united by the bond of their own mutual gravitation towards each other. This should be called a real double star."

We now come to *the* paper * of Herschel's in the ' Philosophical

* "On the Changes which have happened in the Relative Situation of Double Stars."

Transactions,' 1803. This was read on June 9 of that year, and in it he deals with the actual measurements and lays the true foundation of double-star astronomy. Here too, I think, we get an insight into Herschel's character and an explanation of his reticence in connection with binary stars. He evidently, at a very early period, did suspect the real nature of the measures he was so diligently making, but refrained giving an opinion until he could support it by actual facts. Here are his own words :—" That among the multitude of the stars of the heavens there should be many sufficiently near each other to occasion mutual revolution must appear highly probable. But these considerations cannot be admitted in proof of the actual existence of such binary combinations. I shall therefore now proceed to give an account of a series of observations on double stars, comprehending a period of about 25 years, which, if I am not mistaken, will go to prove, that many of them are not merely double in appearance, but must be allowed to be real binary combinations of two stars, intimately held together by the bond of mutual attraction."

He first deals with the observed motion of Castor. He shows that it cannot be explained by the motion of the Sun in space, nor by parallax; but that proper motion, and motion of one star round the other, will both satisfy the conditions of observation: and here he remarks : " As I have now allowed, and even shown, the possibility that the phenomena of the double star Castor may be explained by proper motions, it will appear that, notwithstanding my foregoing arguments in favour of binary systems, it was necessary, on a former occasion *, to express myself in a conditional manner, when, after having announced the contents of this paper, I added, ' should these observations be found sufficiently conclusive,' for, if there should be astronomers who would rather explain the phenomena of a small star appearing to revolve round Castor, by the hypothesis we have last mentioned, they may certainly claim the right of what appears to them the most probable." Proceeding to examine the observed angles more in detail, and making use of the note † of Dr. Bradley's about 1759, he deduces a period of revolution of 342 years. Of course this is only a very rough estimation and made on the assumption of circular motion. Treating each star independently in a similar manner to Castor, he deduced the following periods :—

γ Leonis ...... 1200 years.	ϵ Boötis ...... 1681 years.
δ Serpentis .... 375 ,,	γ Virginis .... 708 ,,

About this period Sir William Herschel seems to have opened up so many branches of astronomical observation that he could give but little time to the double-star section. In particular I

---

* Phil. Trans. 1802, p. 486.

† Double star Castor—no change of position in the two stars, the line joining them at all times of the year, parallel to the line joining Castor and Pollux in the heavens, seen by the naked eye.

cannot but mention my surprise at the extraordinary amount of work he did on variable stars. Ideas seemed very hazy with respect to these phenomena, and the number then known was very small. He reduced observations to a system, observed new variables, formed catalogues, and, in fact, constructed this branch of astronomical research as he had previously done the double-star department. But this is by the way, and I only bring it forward under the impression that Herschel suspected a connection in the two phenomena.

His last paper * on double stars was read before the Royal Astronomical Society, 1821, June 8.

We may regard the quarter-century from 1779 to 1804 as a period by itself, a period during which most important observations were made, both practically and metaphorically, in the dark. Their true value could not be properly assessed at the time, and they were accordingly allowed to lie fallow for some sixteen years.

To introduce the second period, I believe a free use of the address of Francis Baily to be the best course. On the occasion of presenting medals, as President of the Royal Astronomical Society, to John Herschel, South, and W. Struve in 1826, he says :—"The singular and extraordinary changes that had been observed by Sir William Herschel in his reviews of the heavens in 1802 and 1804 had determined Mr. Herschel to follow up the intentions of his father by a review of all the double stars inserted in his catalogues, and as early as 1816 he had commenced this arduous undertaking. Mr. South also being disposed to pursue the same inquiry, suggested the plan of carrying on their observations in concert; and with the aid of two excellent achromatic telescopes belonging to the latter, they employed the years 1821, 1822, and 1823 in this research †.... The remarkable phenomena, first brought to light by Sir William Herschel, have been abundantly confirmed, and many new objects pointed out as worthy the attention of future observers.

" Whilst these important inquiries were carrying on in England, one of our Associates, Professor W. Struve, was engaged in similar observations at Dorpat in Russia. I cannot omit this opportunity of noticing the labours of M. Amici on double stars. With some excellent and beautiful telescopes and micrometers of his own workmanship and construction, this indefatigable and careful observer has extended his examination to upwards of 200 double stars."

The work of Herschel and South is in Part ii. of the 'Philosophical Transactions,' 1824, under the title of "Observations of 380 double and triple stars made 1821, 1822, 1823." Beyond the

---

* "A Catalogue of 145 New Doubles."

† Mr. South was at this time in practice as a surgeon, and the measures were made with refractors of 3¾- and 5-inch apertures at Blackman Street, London. He afterwards set up the 5-inch at Passy, near Paris.

mere measures, there is attached in the form of notes a complete history of each star and a discussion of its motion, if any. The value of this work was acknowledged in France by the presentation of the Lalande Medal to the authors, and by the appearance in the 'Connaisance des Temps' for 1828 of copious abstracts and notes by Arago. The measures made by J. Herschel in 1816 are inserted, and also a very useful correction to Sir W. Herschel's measures. Unless otherwise stated, Sir W. Herschel gave only single measures of angles and distances, and his times are those when the star was first discovered double. In the present work the mean of all his measures is given with the true date of observation, thus rescuing a multitude of unpublished observations, and somewhat atoning for the incorporation of so many wide pairs.

I shall now quote a passage from this work concerning the observation of faint companions. " A rather singular method of obtaining views, and even a rough measure of the angle of stars of the last degree of faintness, has often been resorted to, viz. to direct the eye to another part of the field. In this way, a faint star in the neighbourhood of a large one will often become very conspicuous, so as to bear a certain illumination, which will yet disappear, as if suddenly blotted out, when the eye is turned full upon it." He explains this by supposing the lateral portions of the retina less fatigued by strong lights and less exhausted by perpetual attention, and therefore more sensitive to faint impressions. This method I have myself found useful in observing small planets with the transit-circle, by looking at the wires and so detecting the planet just entering the field.

It should here be mentioned that in W. Herschel's measures he imagined the principal star at the intersection of a meridian and a

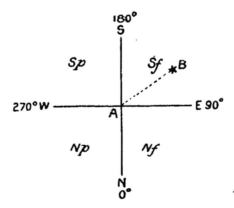

parallel of declination, and the *comes* was referred to one or other of the four quadrants, which were designated as shown in the diagram—north preceding, north following, south preceding, south

11

following,—the angle always being measured from the meridian. Thus he would denote the position of B as 60° *Sf.* This method has given rise to unfortunate mistakes. The method at present in use was adopted by Struve, and consists in making the reference a point (the North Point) and counting the degrees in the angle from the north, through the east, completely round from 0° to 360°. Thus the position-angle of B is NAB, or 120°.

Double-star observers are perhaps more familiar with the name of Struve in this connection than with that of either of the Herschels. This, of course, results from the fact that the numbers used by him in his great catalogue have been retained, in preference to those of Herschel, to denote and tabulate double stars. This adoption obviously arose from the fact that the numbers, in general, run in order of Right Ascension, and also by reason of the great convenience of having reference numbers to so huge a catalogue.

Struve began double-star work at Dorpat as early as 1813 by measuring the differences of Right Ascension and Declination. The result of his work from 1813 to 1820 was a catalogue * of 795 double stars, of which number some 500 were within the limiting distance of 32″. In 1824 he received from Fraunhofer the celebrated Dorpat Refractor; the possibilities thus within his reach determined him to start anew and on a higher plane. Concerning his programme I prefer that Struve should speak for himself, and so quote from his interesting letter to J. F. W. Herschel, published in volume 2 of the Royal Astronomical Society's Memoirs :—

"You may easily imagine with what interest I have perused the work on double stars, and with what pleasure I found that, independently of one another, we have arrived at the same results and deductions. Although my instruments with respect to measurements were formerly † inferior to yours (as I could only observe differences of right ascension in the meridian and angles of position with a 5-feet telescope of Troughton), they may be considered, in an optical point of view, equal to yours ; viz., the 5-feet telescope of Troughton to yours attached to the 5-feet equatorial, and the 8-feet one of Dolland, to yours attached to the 7-feet equatorial, and after receiving the repeating micrometer of Fraunhofer, which I fixed to Troughton's telescope ‡, every desideratum in this instrument was fulfilled; whilst for the angles of position I felt greatly the want of a parallactic apparatus, since I was obliged to determine the zero-point anew for every observation. Nevertheless I succeeded, previous to the arrival of Fraunhofer's refractor, in repeatedly observing the greater part of the double stars included in my catalogue. On receiving, in the autumn of 1824, the Fraunhofer refractor I determined on a new examination of all the double stars observed before, as well as on a minute review of the heavens, from the North Pole to −15° of declination. I

* 'Catalogus 795 stellarum duplicium,' published in 1822.
† 1813 to 1820.   ‡ 1821.

have now accomplished one-third of the labour, and have founa 1000 double stars of the first four classes, among which 800 are new, and of these nearly 300 are of the first class. I shall extend the examination to all stars of the 8th and 8·9 magnitudes ; those of the 9th are not included, as the number would be too considerable. . . . Although the examination of the heavens occupies me almost exclusively, I have, nevertheless, accomplished the absolute determination of the places of all the double stars hitherto observed (including 145 new ones of Sir W. Herschel) by means of the Reichenbach meridional circle. . . . The micrometer measurements are performed generally with a magnifying-power of 480, sometimes with one of 320, for difficult objects and under favourable circumstances with one of 600. The results given are so decisive that the observation of a double star during two evenings (in each of which the distance is measured four times, and the angle of position three times) is sufficient."

The first-fruits of his new endeavours appeared in 1827 under the title of 'Catalogus Novus.' This catalogue contained 3112 double stars arranged in classes according to their distances, with the approximate right-ascensions and declinations, and a rough description. For the formation of this catalogue Struve examined some 120,000 stars, and worked assiduously for 2½ years. This magnificent work formed a worthy forerunner to his *chef-d'œuvre*.

In 1837 he published his great catalogue known as the 'Mensuræ Micrometricæ,' a work justly esteemed by every double-star investigator and observer, but probably an unknown book, visually, to many *. It measures 20 inches by 12, is 1½ inch thick, weighs some 6½ lbs., and therefore does not lend itself very readily to the borrower.

To prepare this great work, containing in all 2640 double stars †, Struve examined ·629 of the total superficial extent of the heavens, and made 11,392 complete sets of measures, a set being the measures of a pair on one night, which, in general, consisted of 3 angles and 4 distances. He followed Herschel's plan of arranging the pairs in classes according to their distances, a considerable drawback to the catalogue, which is, however, somewhat atoned for by a full index arranged in accordance with the Σ number, and giving the page in the catalogue on which the star may be found. There is also an index arranged in Herschel's classes and numbers with the corresponding Σ number, and also one for all the Flamsteed stars. The placing in classes soon leads to complications because of the ever varying distances. It must also be borne in

---

* Lord Lindsay's 'Summary of Struve's Catalogue' contains the mean results of all his observations, and as the stars are arranged in order of R.A. it is more convenient if individual measures are not required. The recently published Memoir of the R. A. S., vol. lvi., is practically Struve's Catalogue brought up to date.

† The stars are numbered up to 3134, but 494 were rejected for various reasons.

mind that the Struve and Herschel classes are not accordant. Struve's eight classes include only stars under 32″ apart, and hence correspond to Herschel's first four classes. The full title of Struve's Catalogue is : 'Stellarum duplicium mensuræ micrometricæ per magnum Fraunhoferi tubum annis a 1824 ad 1837 in Specula Dorpatensi institutæ, adjecta est synopsis observationum de stellis compositis Dorpati annis 1814 ad 1824 per minora instrumenta perfectarum.'

Sir John Herschel had now provided himself with a 20-ft. reflector, the 18-inch mirror of which he had himself ground ; and later he acquired the 7-ft. achromatic used by South. With these instruments he discovered an enormous number of new doubles and made thousands of micrometer measures. With the 20-ft. reflector he compiled the following catalogues :—

1. 321 objects (321 new).... Mem. Ast. Soc. vol. ii. 1826.
2. 295 ,,   (295 ,,)....   ,,   ,,   ,, vol. iii. 1829.
3. 384 ,,   (384 ,,)....   ,,   ,,   ,, vol. iii. 1829.
4. 1263 ,, (937 ,,)....   ,,   ,,   ,, vol. iv. 1831.
5. 2007 ,, (1304 ,,)....   ,,   ,,   ,, vol. vi. 1833.
6. 285 ,,   (105 ,,)....   ,,   ,,   ,, vol. ix. 1836.

Making altogether 4555 multiple stars, of which 3346 were new.

Amongst the notes in the first catalogue published in 1825, I find the following answer to a pertinent question frequently asked in other departments of astronomy, and Double-Star Observers should act on the hint conveyed in the following lines :—" It may be enquired why we should aim at increasing our list of double stars already so numerous, and why this list should be carried down to such minute objects. To such a question I apprehend the answer may be found in this consideration : the labour of the Astronomer is much like that of one who should examine, grain by grain, the sands of the Sea, in the certainty that among them numerous grains must exist of extraordinary value, or of singular properties. The more individuals our search embraces of a class which has already proved productive, the greater our chance of further success ; and so long as no presumption *à priori* can be adduced why the most minute star in the heavens should not give us that very information respecting parallax, proper motion, and an infinity of other interesting points which we are in search of, and yet may never obtain from its brighter rivals, the minuteness of an object is no reason for neglecting its examination. But if small stars are to be watched, it is necessary they should be known; nor need we fear that the list will become overwhelming. It will be curtailed at one end, by the rejection of uninteresting and uninstructive objects, as fast as it is increased by new candidates."

The measures in these catalogues were made during the examination of star-clusters and nebulæ which Herschel commenced in

1825 and completed in 1832, and which he considered his *magnum opus.*

He made two sets of measures with the 7-ft. achromatic :—

1. Measures of 364 double stars, 1828–30 .. Mem. R. A. S. vol. v.
2. A second series, 1831–33 ....   „   „   vol. viii.

In 1833 he left England for the Cape of Good Hope, and there in the years 1835–38 he examined 4000 nebulæ and star-groups and nearly 3000 double stars. They were not published until 1847; but the delay is quite accounted for by the fact that the whole mass of material was arranged, reduced, and prepared for press entirely by himself.

Sir J. Herschel was preceded in the South by Mr. James Dunlop. This pioneer had gone out to Paramatta with Sir Thomas Brisbane in 1823, and when the latter returned in 1827 Dunlop resolved to remain for the purpose of making a general survey of the heavens from the South Pole to 30° south declination. He had at his disposal a small achromatic telescope of 46-inch focus, a 9-ft. reflector, a parallel wire micrometer, and an Amici double-image micrometer. Amongst his other work he completed a catalogue of 253 double stars with accompanying micrometer measures *.

When Herschel left England the number of double-star observers was fast increasing. Two, at least, of these demand especial notice. The Rev. W. Dawes commenced double-star observing at Ormskirk in 1830 with a 3·8-inch refractor, and the work which he proposed doing is given in the following extract from his communication † to the Royal Astronomical Society :— " Last autumn I commenced a review of all the double stars within the distance 16", which were measured as new, or, for the first time, by yourself ‡ and Sir J. South .... and am pleased to find, contrary to expectation, that I do not, in any instance, experience more difficulty in procuring the observations than Sir James appears to have encountered .with his 7-ft. equatorial." This physical characteristic has earned for him the appellation of "the eagle-eyed Dawes." Sir G. Airy, in presenting to him the gold medal of the Royal Astronomical Society in 1855, remarked : " Distinguished as Mr. Dawes has been by an extraordinary acuteness of vision, and by a habitual, and (I may say) contemplative precision in the use of his instruments, his observations have commanded a degree of respect which has not often been obtained by the productions of larger instruments." From 1839 to 1844 he was at Mr. Bishop's Observatory and there had the use of a 7-inch refractor. He afterwards had a 6-inch Merz, which gave place to a fine $7\frac{1}{2}$-inch Alvan Clark, and later to an $8\frac{1}{4}$-inch by the same maker. These telescopes became famous for their performance in the hands of Dawes. His work is scattered

* Mem. R. A. S. vol. iii. 1829.
† Mem. R. A. S. vol. v. p. 141.
‡ Sir J. Herschel.

throughout the 'Monthly Notices' and the 'Memoirs' of the Royal Astronomical Society. The following are the principal :—

121 Double stars, 1830–1833, Ast. Mem. vol. viii.
100   „   „   1834–1839, „   „   vol. xix.
250   „   „   1839–1844, Mr. Bishop's Volumes.

·The great work of Dawes is in the Astronomical Society's Memoirs, vol. xxxv., and is entitled "A Catalogue of Micrometrical Measures of Double Stars." Copious notes on each pair are appended, and in addition are some valuable remarks on the parallel wire, the spherical crystal, the prismatic crystal, the Amici double image, and the four glass double-image micrometers, and also some very practical hints on observing and on telescopic apertures. These remarks are published in the 'Monthly Notices,' 1867, April, and are worth more than ordinary attention.

I was particularly pleased when I came upon the following paragraph :—" A curious fact with which I have been familiar for more than 30 years, may perhaps be worthy of notice in this place : namely, that stars at small altitudes require a shorter focus than those at large altitudes, to be seen with perfect distinctness. Of course the difference is slight, yet it is decided and constant. It is independent of the brightness of the object, but yet is, I think, most obvious when the actual difference of magnitude is just so far in favour of the lower star as to render its apparent brilliancy equal to that of the higher."

Admiral W. H. Smyth commenced observing at Bedford also in 1830 with an equatorially mounted refractor of 5·9 inches aperture, an instrument considered one of Tulley's best works. The work performed here in the years 1830 to 1839 has been embodied in the two familiar works known as 'A Cycle of Celestial Objects' and the 'Bedford Catalogue.' The descriptions of the various objects are justly popular and form anything but dull reading. The 'Cycle' contains positions, micrometrical measures, and full particulars of

419 Double Stars,   98 Nebulæ,
 20 Binaries,   72 Clusters,
 46 Quadruples,   161 Stars and Comites,
 21 Multiples,

for the epoch 1840. It was published in 1844, and he received for it the medal of the Royal Astronomical Society. The 'Bedford Catalogue' contains 850 celestial objects arranged in order of Right Ascension, and includes all his micrometrical measures. Smyth afterwards transferred his telescope to Dr. Lee's Observatory at Hartwell, and his later observations are published in the 'Ædes Hartwellianæ' and in more complete form in a work which appeared in 1860 entitled 'Speculum Hartwellianum.'

During this period double-star astronomy was not neglected on the Continent. As Sir John Herschel took possession of double-star work in succession to his father, so in like manner the mantle of W. Struve seems to have fallen upon his son Otto Struve, and also upon Maedler his successor in the Directorship of the Dorpât Observatory. Maedler had for many years been interested in problems relating to the construction of the Universe. Consequently, when W. Struve removed in 1840 to Poulkowa, it seemed natural that Maedler should endeavour to maintain the high reputation of his Observatory by continuing Struve's work. And most eminently did he succeed. Each volume of the 'Dorpât Observations' is replete with micrometric measures, notes, and discussions of the motion of binary stars. The outcome of his early work (1834–45) is perhaps his best claim for recognition. I refer to his 'Untersuchungen über die Fixstern-Systeme,' published in 1847, a work abounding in measures, catalogues, and discussions, which will be treated of later *. Maedler continued observing up to 1861.

Otto Struve succeeded his father at Poulkowa in 1858, but his double-star work dates from a much earlier period. Already in 1843 he had published his catalogue of 514 double and multiple stars. This "Poulkowa Catalogue" of O.Σ stars is a small one, but fully as important as any preceding it, containing a larger proportion of close and interesting pairs. The revised edition of 1850, purged of errors, contains 424 stars, whose components are under 16″ distance. The following table will show in a marked degree the relative importance of the three catalogues of northern stars published up to 1850. Of course this table is not to be taken as any disparagement of the earlier works. O. Struve's catalogue was compiled as a result of a study of these, with superior instrumental power, and with higher knowledge of the phenomena involved :—

Distances	0″ to 1″	1″ to 2″	2″ to 4″	4″ to 8″
W. Herschel	9	20	45	49
W. Struve	38	82	128	112
O. Struve	158	64	53	54

The great mass of O. Struve's measures made in the years 1839–75 are published in Vols. ix. and x. of the "Poulkowa Observations." The Poulkowa telescope used by O. Struve is a fine 14·93-inch Merz of 22·55 ft. focus.

But double-star work was not confined to the more northern latitudes. In the clear skies of Italy two observers were amassing

* The investigations of the motion of double stars will be treated of separately.

material for building the edifice whose foundations were laid by the Herschels and Struves.

Baron Dembowski seems to have imbibed his love of astronomy in his nautical career, and it is probable that his choice of work depended in a great manner upon the suggestions of Don Antonio Nobile, of the Observatory "Capodimonte." In 1851 Dembowski acquired a telescope of 5 inches aperture, and at once set to work on his programme, which was no other than the revision of the brighter pairs in the 'Measuræ Micrometricæ.' The amount of energy this instrument absorbed only Dembowski knew. Here is what O. Struve and Schiaparelli said of it :—" Quel Deàlite era certamente di una grande perfezione ottica, rispetto all sue dimensioni, e all' occhio acuto ed exercitato di Dembowski permetteva di separare distintamenti coppie di stelle distanti fra loro meno di un secondo di arco. Ma d' altra parte la contruzione meccanica lasciava molto a desiderare. Esso non aveva nè moto d'orlogio, e neppure un circolo di posizione." A bad driving-clock with a good setting-circle may be endured; but no driving-clock and no setting-circle—well, Dembowski under these conditions turned out work justly esteemed. This state of things, however, in eight years proved too much even for Dembowski, and he ordered a fine 7-inch Merz with a good clock movement. With this he set himself to measure all the stars in the Dorpât Catalogue and all those in the Poulkowa Catalogue which he could reach. This programme was afterwards greatly extended, as will be noticed from the contents of the two volumes * of his measures edited by O. Struve and Schiaparelli.

When sending in 1869 his last batch of measures to the 'Astronomische Nachrichten,' completing his work on the Struve Stars, he wrote :—" For various reasons he had been unable to satisfactorily measure 64 of the stars in Struve's Catalogue. This result is the highest praise that can be given of the excellence of the Merz object-glass." Dr. Huggins, when presenting the Royal Astronomical Society's medal to him in 1878, referring to this says : " May we not even more justly say that it is at least evidence as remarkable of the skill and zeal of the observer ? "

Secchi made a considerable number of measures in the two periods 1856-59 and 1863-66. These measures are in two distinct catalogues, and will be found in ' Memorie dell' Osservatorio del Collegio Romano.'

We have now fully considered the work of the principal observers in what we may term the second period of Double-Star Astronomy, a period when the workers were more or less feeling their way. There were many others, e.g. :—

---

* 'Misure Micrometriche di Stelle Doppie e Multiple fatte negli anni 1852-1878 dal Barone Ercole Dembowski.' Roma, 1883.

Bessel	{ whose measures extend from } 1814–34	Konigsberg *.
Encke & Galle	,, 1838–48	Berlin.
Challis & Glaisher	,, 1839–42	Cambridge.
Kaiser	,, 1840–66	Leiden.
Simms, Philpott, Morton	,, 1843–59	Ld. Wrottesley's Obs.
Jacob	,, 1845–58	Poonah.
Mitchel	,, 1846–48	Cincinnati.
Fletcher	,, 1850–53	Tarn Bank.
Gilliss	,, 1850–52	S. America.
Powell	,, 1853–62	Poonah.
Winnecke	,, 1855–56	Altona.
Alvan Clark	,, 1857–60	N. America.
Knott	,, 1860–77	Cuckfield.
Romberg & Talmage	,, 1862–72	Leyton.
Main	,, 1862–77	Oxford.
Engelmann	,, 1864–86	Leipzig.
Ellery	,, 1866–70	Melbourne.
Wilson & Gledhill	,, 1872–80	Bermerside.

The third period began when Burnham, who had been working at Chicago with a 6-inch refractor mounted in his back-yard, sent a modest catalogue of 81 new double stars to the Royal Astronomical Society in March 1873. This catalogue was followed by others, so that by 1882 his number of new pairs exceeded the thousand. At this time Burnham was working with the $18\frac{1}{2}$-inch Dearborn refractor, and in the preface to his 1882 catalogue he remarks : "The present catalogue will conclude my astronomical work, at least so far as any regular or systematic observations are concerned." We know that fate willed otherwise, and Burnham did not give up active observing until he was sure his work would be continued by such enthusiasts as Professors Hussey and Aitken. Burnham's experience of telescopes is unique; with the

6-inch, at Private Observatory, he discovered	454 pairs.
$18\frac{1}{2}$,, Dearborn ,, ,,	418 ,,
36 ,, Lick ,, ,,	201 ,,
$15\frac{1}{2}$,, Washburn ,, ,,	88 ,,
12 ,, Lick ,, ,,	59 ,,
40 ,, Yerkes ,, ,,	56 ,,
$9\frac{1}{2}$,, Dartmouth Coll. Observ., ,,	29 ,,
26 ,, Washington Observ., ,,	15 ,,
16 ,, the Warner ,, ,,	2 ,,

There is no need to refer to the separate catalogues, as Burnham, in 1900, collected all into one "General Catalogue of 1290 Double Stars discovered from 1871 to 1899 by S. W. Burnham." It was published as the First Volume of the Yerkes Observatory Publications. He arranged all the stars in order of R.A., retaining the original numbering, and giving all the measures made by himself and by other observers. Diagrams and notes make it a very complete and compact work.

* Vol. 35, Konigsberg Observations, contains the measures made by Bessel, Schlüter, Peters, Luther, and Auwers from 1830–62, a very important series.

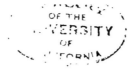

As already stated, Hussey and Aitken took up Burnham's work, and with the 36-inch Lick refractor have added an immense number of close pairs, from which we may hope to find a large percentage of binaries. An analysis of the discoveries of these three observers will afford an idea of the advance in this third period :—

	Under 0″·5.	0″·5–1″·0.	1″·0–2″·0.	2″·0–3″·0.	3″·0–6″·0.
Burnham	125	246	344	175	205
Hussey	369	306	324	157	201
Aitken	518	347	387	199	269

To make regular measures of these and of the older pairs is, if less exciting, an absolute necessity if the discoveries are to be of any use in increasing our knowledge of stellar systems. The three already mentioned have contributed to regular measurements *.

Asaph Hall, with the Washington 26-inch refractor, made annually a number of measures which appeared in the Observatory Volumes, and were afterwards collected into two volumes :

(1) Measures of 423 stars made from 1875 to 1880.
(2) Observations of double stars, 1880–1891.

Seabroke has a long series of measures (1876–1905), in the Mem. R. A. S., made at Rugby with an 8¼-inch Clark.

The Cincinnati Publications contain over 4300 measures of stars between the Equator and 30° S. Dec. (1875–1880).

Dunér, working with a 9·8-inch at Lund from 1867–1876, made 2679 observations of stars selected from the Struve and Otto Struve Catalogues. These measures, together with those of former observers, he published in 'Mesures Micrométriques d'Étoiles doubles," 1876.

G. W. Hough at Dearborn Observatory observed with the 18½-inch refractor with which Clark discovered the faint companion to Sirius. His discoveries—630 pairs—and his measures are scattered throughout the *Astronomische Nachrichten*; but quite recently Prof. Doolittle has collected them, and, adding his own measures, published them as a volume of the Flower Observatory.

Doberck, who has done so much for double-star astronomy, commenced his measures at Markree with a 13·2-inch Cauchoix

---

* Lick Observatory, vol. ii.; Yerkes, vol. i.; Mem. R. A. S. vols. xliv., xlvii.; and 'General Catalogue' : for Burnham. Lick Bulletins and Lick Observatory vol. v. for Hussey and Aitken.

refraetor. Afterwards, at Hongkong, be continued his work with the Lee equatorial. His measures will be found in the *A. N.* nos. 2196-7-8-9, 2242, 2989, 3023, 3378, and 3466. There are some in the Trans. R. Irish Academy, vol. xxix.

Maw started measuring with a 6-inch Cooke at Kensington in 1888. Afterwards in 1896 he procured, and set up at Outwood, the 8-inch of the late George Hunt, an instrument formerly belonging to Dawes. His measures will be found in the various volumes of the Memoirs of the R. A. S., commencing with vol. l.

In 1893 the Astronomer Royal, Sir W. Christie, decided that double-star work should form a portion of the regular routine at Greenwich. Measuring was started, with the 12¾-inch Merz, by Lewis, who continued the work in 1894 with the 28-inch refractor. Other observers have since shared in the work, viz. Bryant, Bowyer, and Furner.

M. Bigourdan, at the Paris Observatory, devoted much time and energy both in actual observation, and investigations of personal errors, forms of micrometer, and other points. His results will be found in the Paris Observatory Publications.

These are but a few of many equally enthusiastic double-star observers, a complete list of whom may be found in the Introduction to vol. lvi. Memoirs R. A. S. In this list we find :—

Knott	7¼-inch, formerly belonging to Dawes. Memoirs R. A. S. xliii.
Tarrant	10¼-inch reflector. Ast. Nach.
Perrotin	30-inch Nice refractor. Nice Observatory Publications.
H. Struve	30-inch (Poulkova). Poulkova Publications.
Glasenapp	6½- and 9½-inch. Measures made at St. Petersburg and Domkino.
Leavenworth ...	10-inch (Haverford), 16-inch (Northfield), 26-inch (McCormick).
See	24-inch (Lowell), 26-inch (McCormick).
Scott	5-inch (Shanghai). B. A. A. Journal.
Espin	17¼-inch reflector. M. N. Roy. Ast. Soc.
Kustner	9½-inch (Hamburg). Ast. Nach.
Biesbroeck	12-inch (Heidelberg), 15-inch (Uccle). Ast. Nach. and Annals Brussels Observatory.
Schiaparelli......	8-inch (Brera), 18-inch (Brera). Pub. de Reale Osserv. di Brera in Milan, xxxiii.
Lau	10-inch (Copenhagen). Ast. Nach.
Doolittle	18-inch (Flower). Publications Flower Observatory.
Comstock	15½-inch (Washburn). Publications Washburn Observatory.
Coleman	8-inch (Cooke). Memoirs R. A. S.

This scattering of discoveries and measures throughout a great variety of observatory publications and magazines made it difficult to form a proper working catalogue, and an enormous labour if material is required for investigations. Admiral Smyth's 'Celestial Cycle,' already referred to, soon became of little use for the purpose. In 1874 the Royal Astronomical Society published, in vol. xl. of their ' Memoirs,' "A Catalogue of 10,300 Multiple and Double Stars, arranged in Order of R.A. by the late Sir J. Herschel, Bart." But this was a catalogue only—no angles and no distances; it served only to bring together the known pairs. In 1878 Flammarion's 'Étoiles Doubles' came as a boon. It contained 819 systems, and M. Flammarion had collected from various sources some 14,000 measures of these. In obtaining recent measures he had the assistance of Gledhill, Wilson, Seabroke, Burnham, and Dembowski. A large number of measures he made himself with a defective 15-inch stopped down to 7 or 8 inches. The stars are arranged in order of R.A., with the R.A. and N.P.D. for 1880, the star's name, the measures, and notes. A most compact work.

This splendid attempt to cope with the needs of this branch of astronomical research was followed, in 1879, by Crossley, Gledhill, and Wilson's 'Handbook of Double Stars.' This book is divided into four sections :—1. Methods and instruments; 2. Calculation of an orbit; 3. Measures and notes of 786 stars—the measures not in chronological order, but arranged under each observer; 4. Titles of works on double stars and on double-star colours.

Burnham's Catalogue, already described, of his own discoveries was the next serious attempt. This was followed by part ii. vol. ii. of the 'Annals of the Royal Observatory, Cape of Good Hope,' "A Reference Catalogue of Southern Double Stars " by R. T. Innes. Double-star work in the southern hemisphere has always been of a desultory kind, and this Catalogue was a thing long desired. The ordinary southern observer had three courses open to him :—

1. He could search for new pairs.
2. He could measure pairs near the equator which were already amply looked after by northern observers.
3. After spending an inordinate time searching pamphlets, he could make out a poor observing list.

Innes does this once for all. He arranges his 2140 pairs in order of R.A., and gives all necessary measures and notes. There are other features of this volume to which reference will be made later. The character of the stars is as follows :—21 per cent. are under $1''$ separation; 24 per cent. are between $1''$ and $2''$. Southern observers owe a great debt to Mr. Innes, and also to Sir David Gill for obtaining its publication.

Prof. Hussey in 1901 published, as vol. v. of the 'Lick Observatory Publications,' his " Micrometrical Observations of the Double Stars discovered at Poulkova, together with the Mean Results of the Previous Observers of those Stars." A most valuable work, in which Prof. Hussey does for the Otto Struve stars what Burnham had already done for the Burnham stars.

The only really important series left was the stars contained in Struve's 'Mensuræ Micrometricæ,' to the number of 2640. In 1906 these were dealt with by Lewis in vol. lvi. of the 'Memoirs of the Royal Astronomical Society.' And to crown all came, in 1907, the 'General Catalogue' of Prof. Burnham, which consists of two volumes :—

Vol. I. The Catalogue of 13,665 double stars
„ II. Notes, measures, and references to most of the stars in Vol. I.

Thus within a period of eight years practically the whole department of double-star astronomy has been overhauled, remodelled, and given a new start. Would-be observers should find no difficulty in determining what to observe ; there is plenty to do, and there is work suitable to any instrument, from the smallest to the 40-inch refractor at Yerkes.

In passing so lightly over the labours which have brought about this satisfactory state, I feel an injustice is done ; but opportunities for writing more fully on points peculiar to each work will be afforded.

We may now pass to the consideration of " What is a double star ? " We have seen that in the earlier stages many wide pairs were measured and catalogued which would not be tolerated now. But when once they have been observed and catalogued, it is difficult to see how we can discard them. Could we make sure that the attention given to such pairs would in no way prejudice the measurement of closer pairs, then let us retain them. Indeed, the altered circumstances of recent years enable us to do so with a certain degree of safety, for we may keep them on our lists and look for measures from the Astrographic plates. This does not, however, mean advocating the compilation of new lists of wide pairs by double-star observers, as no doubt these will come automatically from workers in the astrographic branch. These ideas will be found in practical agreement with other observers, for instance :—

W. Struve rejected pairs whose separation exceeded 32".

O. Struve placed the limit for retention at 16".

O. Stone rejected pairs whose separation exceeded 32" where the magnitude of the principal star was fainter than 8, and, further, if fainter than 9th mag. the limit was reduced to 16", unless the pair was included in the formula

$$m + m' = 19 - \frac{D}{5},$$

which discriminated against faint companions.

Innes, in compiling his 'Reference Catalogue,' adopted the following scale :—

Mag. of primary.	Limit of separation.	Mag. of primary.	Limit of separation.
1	30″	6	7″
2	25	7	5
3	20	8	3
4	15	9	1
5	10		

These and other criteria are useful as rough guides, but must not be strictly adhered to. Stone and Innes certainly regarded them as such. In fact, if we applied the criteria of Innes to pairs well known we should exclude :—

Σ 518.—A = 4·5, B = 9·2, C = 10·9, where BC is a binary of about 180 years' period, whose separation has varied from 4″·2 to 2″·2, Innes' limit being 1″. The system BC revolves about A at a distance of 82″ to 89″, Innes' limit being 12″.

Σ 2220.—A = 3·8, B = 9·5, C = 10·5, where BC is a binary of about 43 years' period, whose separation at discovery was 1″·8. BC revolves about A at 31″ dist.

OΣ 547.—A = 8·6, B = 8·8, and dist. = 4″·6. A very interesting pair, with a proper motion of 0″·9 in direction 98°.

Others could doubtless be found, such as 61 Cygni and Σ 2481, and it is apparent that we cannot fix limits. Aitken and Hussey rarely keep a pair separated over 6″.

Writing of limiting separation brings to mind the question often asked as to the separating power of telescopes. Theoretically this is not difficult to answer; but in most cases it is not the separating power that is really meant. What the enquirer usually seeks for is rather what we may call the observing or detecting power—the power to discriminate between a single and a double star. The separating power is given with sufficient accuracy by the formula

$$\text{Separating Power} = \frac{4''·56}{a},$$

where a is the aperture of the O.G. in inches.

Thus the separating power of the Lick 36-inch refractor is 0″·13, and for the Greenwich 28-inch Grubb 0″·17, and when the seeing is good the images of the component stars have been seen distinctly separated at these limits. However, to return to the main question, the answer to which involves

1. Size of objective;
2. Quality of objective;
3. Condition of atmosphere;
4. Personality;

or just the same set of conditions as govern "range of visibility," or the ability to see faint objects. It will be seen, therefore, that a direct answer is impossible, but we have a means of knowing what has been done. Reference has been made to Burnham's unique experience with various telescopes, and fortunately he has made discoveries with all. Now if Burnham discovers a pair with a small telescope, the reality of the pair admits at once of direct proof by turning on to it a telescope of higher power. Bearing this in mind, at a time when there was some misconception on the point, the closest pairs discovered were tabulated under each instrument with the following result :—

Aperture in inches.	Smallest separation.	Mean sep. of three closest.
6	0″.30	0″.32
9.4	0 .28	0 .37
12	0 .27	0 .30
15.5	0 .31	0 .36
18.5	0 .20	0 .25
36	0 .13	0 .14

These stars have since been repeatedly measured and some have proved binary. Their reality is not open to question. Burnham is an exceptional observer, but we must remember that to detect is more difficult than to verify. Incidentally one could infer the quality of the object-glass.

With respect to visibility it would seem that the quality and size of O.G. should override everything else; but sometimes the sensitiveness of the eye and the condition of the atmosphere, in some mysterious way, become the governing factors. Burnham has stated the case clearly thus :—" An object-glass of 6 inches one night will show the companion to Sirius perfectly; on the next night, just as good in every respect, so far as one can tell with the unaided eye, the largest telescope in the world will show no more trace of the small star than if it had been blotted out of existence."

A few words about photography. As far back as 1854, G. P. Bond, of Harvard, pointed out the immense advantage of securing permanent images of double stars, but owing to a want of sensitiveness in his plates and of a good driving-clock he was unable to obtain one till 1857. In that year an exposure of 8 seconds obtained for him on a collodion plate measurable images of ζ Ursæ Majoris. The measurement of 62 impressions gave a distance of $14''.21 \pm 0''.013$. Pickering afterwards made some attempts, and Gould obtained some between 1870 and 1882 at Cordoba. In 1886 MM. Henry succeeded in securing, amongst others, the following pairs :—

γ Virginis	mags. 3.0 and 3.2, distance	5″.34
α Herculis	„ 3.5 „ 5.5, „	4 .73
δ Serpentis	„ 4.0 „ 5.0, „	3 .45
ε² Lyræ	„ 5.7 „ 6.0, „	2 .34

Some experiments were made in this direction in 1894 with the Astrographic Telescope at Greenwich, using a concave triple-lens (cemented), which enlarged the image about fourteen times. The following stars were measured :—

γ Leonis .... mags. 2·0 and 3·5, distance 2″·4
ξ Ursæ Maj....... „   4·0 „ 5·0,   „   1 ·4
λ Ophiuchi ...... „   4·0 „ 6·0,   „   1 ·3

In 1897 some photographs were taken in the principal focus of the 28-inch refractor at Greenwich. Some good results were obtained on dry collodion plates.

	Mags.	Diameters of		Space between images.	Distance.	Distance minus space.	Exposure.
		Large star.	Small star.				
γ Leonis.........	2·0, 3·5	″1·41	″0·88	″1·35	″2·50	″1·15	s 5
Σ 2727	4·0, 5·0	1·32	1·12	9·83	11·05	1·22	4
ξ Ursæ Maj. ...	4·0, 5·0	1·06	0·83	1·02	1·96	0·94	6
Σ 2382	4·6, 6·3	1·47	1·01	1·73	3·04	1·31	6
Σ 2383	4·9, 5·2	1·46	0·93	1·10	2·33	1·23	6
Σ 2130	5·0, 5·1	1·80	1·48	0·59	2·23	1·64	2

These photographs were so taken that each plate had a number of images of the double star, each image resulting from exposures ranging from 1 to 30 seconds. The above measures were made from the best images on the plate. This distance — space is the sum of the two semi-diameters of the stars, and its mean, 1″·25, may be taken to roughly represent the separation at which the two images will just touch. It would, seem therefore that with these particular plates, and an exposure of 5ˢ or 6ˢ, we should obtain photographs of pairs separated 1″·25 *, in which the images shall just touch, provided the photographic action did not spread when the two images approached.

Thiele †, using a negative lens in front of the focus, took 180 plates of 140 double stars. Of these, 125 plates gave measurable images of 110 stars. The separation of the components ranged from 1″·1 to 8″·5. The photographic diameter of principal star ranged from 1″·5 to 6″·4, giving 3″·0 separation for touching images.

* 1″·25 is equivalent to 0·0021 inch on the plates.
† Ast. Nach. 4224.

These are interesting and in some special cases useful; but for general work it will be seen that the photographic method has its limitations, and can never replace the visual method.

As we have seen, Sir W. Herschel did not profess to make any attempt to investigate, in an accurate manner, the orbits of those stars which he considered binary. The honour of being the first to introduce a system of formulæ to calculate the orbit of a binary star belongs to M. Savary, who, in 1827, published his method of computing an orbit from four complete observations. He deduces, in a very neat manner, equations involving the elements of the apparent ellipse, from a consideration of the areas of the elliptic sectors and segments.

If 1 is a point on the apparent ellipse at time t_1,

2	,,	,,	,,	,,	t_2,
3	,,	,,	,,	,,	t_3,
4	,,	,,	,,	,,	t_4,

o the place of the stationary star,

and k a constant area,

then kt is the elliptic sector described in the time t.

Again, if (12) represent the chord between 1 and 2, the area of the segment is $k(t_2 - t_1) - (012)$.

Putting A and B for the semi-axis major and minor of the apparent ellipse, and E_1, E_2, E_3, E_4 for the eccentric anomalies at the points 1, 2, 3, 4.

Then $\dfrac{AB}{2}[(E_2 - E_1) - \sin(E_2 - E_1)]$ is the segment.

For brevity let

$$2\phi = \tfrac{1}{2}[(E_2 - E_1) - \sin(E_2 - E_1)],$$

and then

$$k(t_2 - t_1) - (012) = AB(2\phi)\ \text{*.}$$

By a system of parallel chords the whole set of formulæ follow.

The method is thus founded on the law of gravitation and is purely analytical. A full account is published in the ' Connaissance des Temps,' 1830, and is accompanied by an application to the determination of the orbit of ξ Ursæ Majoris.

Savary also shows how to determine the max. and min. limits of parallax by a combination of the motion of the comes and of light.

In 1832 Encke † gave another method, also based on the knowledge of four complete observations. Encke adheres more closely to the orthodox methods. He applied his formula to determine the orbit of 70 Ophiuchi.

I do not propose entering further into the details of these

---

* This formula is used by Thiele in his graphical process.

† 'Berlin Jahrbuch,' 1832.

methods. They are ingenious and worth perusal by the student; but in practice the formulæ are somewhat cumbersome and tedious.

In 1832 Sir John Herschel communicated to the Royal Astronomical Society a method* "which for elegance and practical utility must, I think, be placed above every other that has appeared." For this " Investigation of the Orbits of Revolving Double Stars " †, he received, 1833, the gold medal of the Royal Society. Before proceeding with his method, a description of the means he employed to make the most of his material will be of advantage. He rejects entirely measures of distance, and depends only on position-angles, any number of which may be used.

In his own words : " It is clearly a mere waste of time to attempt to deal, by any refined or intricate process of calculation, with data so uncertain and irregular—in fact, so excessively loose and insecure are all the measures of distance which we actually possess, that I have no hesitation in declaring that they must, one and all, be peremptorily excluded from any share of consideration in the investigation of the elements, the value of *a* only excepted."

Starting with this idea, Herschel collects all the observations of angle and plots them on cross-ruled paper, taking the dates as ordinates and the measured angles as abscissæ. A smooth curve is next drawn through the dots, and so, by means of the angles before and after any epoch, he determines the angle at the epoch more accurately than can be done by the actual observation. All observations, too, have a certain share in the curve, and wild observations are at once seen isolated.

The next process is to read off the interpolating curve the times corresponding to say every 5 degrees of angle, and take the first differences, and these, divided by the change in the position-angle θ, give approximately a series of values of $\frac{dt}{d\theta}$.

But by the theory of elliptic motion, if r represents the distance,

$$r^2 . \frac{d\theta}{dt} \text{ is a constant;}$$

$$\therefore \quad r = k . \sqrt{\frac{dt}{d\theta}}.$$

So that a series of *relative* values of the distance corresponding to the angles is deduced from the angles.

Thus in place of four positions, Herschel had choice of a large number, which, having laid off, gave him points through which he drew the best possible ellipse. The ellipse then enabled him to correct his interpolating curves, and with new values of $\frac{dt}{d\theta}$ deduce

---

* Airy.   † Mem. Roy. Ast. Soc. vol. v. p. 171.

new normal places. These he again laid off and drew amongst them his final ellipse.

Undoubtedly the distance-measures at Herschel's command were very poor, and his plan of deducing relative distances from interpolating curves was the one most suitable. Originally this was the plan adopted by me; but gradually it was dropped in favour of plotting the angles and distances straightway, and fitting amongst the points the most suitable ellipse. Looking back, it seems to me that the determining factors for discarding Herschel's plan were the difficulties experienced when the angular velocity varied rapidly, and also the increased reliance on distance-measures.

Each computer will please himself in the method of preparing his material for constructing the apparent ellipse. Construct the apparent ellipse he must. Whatever procedure he adopts afterwards, we are all one with Sir J. Herschel in considering the apparent ellipse essential. No pains should be spared in this, as otherwise subsequent work will be vitiated, and time spent on it thrown away.

The area described in the real ellipse is proportional to the time, and these areas when projected still remain in the same relative proportion, and are consequently also proportional to the time. No apparent orbit should be finally adopted until this condition is satisfied as nearly as possible. Ways of doing this will suggest themselves.

Having the apparent ellipse, we can proceed to Sir J. Herschel's method of finding the elements of the true orbit.

Let $N \, \Omega$ be the line of nodes;
$\qquad \Omega ABN$ a portion of the true orbit, C its centre;
$\qquad \Omega A_1 B_1 N$ a portion of apparent orbit, C its centre;
$\qquad$ S be principal star in true ellipse;
$\qquad$ S_1 be principal star in apparent ellipse;
$\qquad$ A the comes, in true ellipse, at periastron.

Then $\quad$ CSA is the semi-axis major of real ellipse $= a$,

and $\qquad$ $CS_1 A_1$ its projection in the apparent ellipse $= a_1$.

The ratio

$$\frac{CS}{CA} = \frac{CS_1}{CA_1} = e, \text{ the eccentricity,}$$

Let $\quad OC \, \Omega$, the angle of node in apparent orbit, $= \Omega$;
$\qquad OCA_1$, the angle of projected periastron, $= \alpha$;
$\qquad OCB_1$, the angle of projected minor axis, $= \beta$;
$\qquad \Omega CA$, in the true ellipse, between the node and periastron, $= \lambda$.

Then in true ellipse draw AR perpendicular to $C \, \Omega$, and join $A_1 R$ in the apparent ellipse.

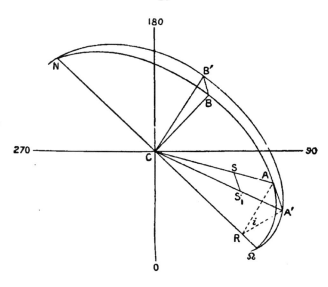

$ARA_1 = i$, the angle of inclination of the orbits.

$$CR = CA \cdot \cos RCA$$

$$CR = CA_1 \cdot \cos RCA_1$$

$$\therefore \quad a \cos \lambda = a_1 \cos (a - \Omega). \quad \ldots \ldots \quad (1)$$

$$AR = CA \cdot \sin RCA$$

$$A_1 R = CA_1 \cdot \sin RCA_1$$

$$\therefore \quad a \sin \lambda \cdot \cos i = a_1 \sin (a - \Omega). \ldots \ldots \quad (2)$$

Similarly

$$- b \cdot \sin \lambda = b_1 \cos (\beta - \Omega) \quad \ldots \ldots \quad (3)$$

$$b \cdot \cos \lambda \cos i = b_1 \sin (\beta - \Omega) \quad \ldots \ldots \quad (4)$$

$\dfrac{(2)}{(1)}$ gives

$$\cos i = \tan (a - \Omega) \cdot \cot \lambda \quad \ldots \ldots \quad (5)$$

$(1)(3)$ gives

$$ab \sin \lambda \cdot \cos \lambda = a_1 b_1 \cos (a - \Omega) \cdot \cos (\beta - \Omega)$$

$(2)(4)$ gives

$$ab \sin \lambda \cdot \cos \lambda \cdot \cos^2 i = a_1 b_1 \sin (a - \Omega) \cdot \sin (\beta - \Omega)$$

$$\cos i = \sqrt{\{- \tan (a - \Omega) \cdot \tan (\beta - \Omega)\}} \quad \ldots \quad (6)$$

$\therefore$ from (5)

$$\tan \lambda = \sqrt{- \frac{\tan (a - \Omega)}{\tan (\beta - \Omega)}}. \quad \ldots \ldots \quad (7)$$

(1)(2) gives

$$a^2 \sin \lambda . \cos \lambda . \cos i = a_1^2 \sin(\alpha - \Omega) \cos(\alpha - \Omega)$$

(3)(4) gives

$$b^2 \sin \lambda . \cos \lambda . \cos i = b_1^2 \sin(\beta - \Omega) . \cos(\beta - \Omega)$$

$$\frac{a^2}{a_1^2} . \frac{b_1^2}{b^2} = -\frac{\sin 2(\alpha - \Omega)}{\sin 2(\beta - \Omega)} = -\frac{\sin 2\alpha - \cos 2\alpha . \tan 2\Omega}{\sin 2\beta - \cos 2\beta . \tan 2\Omega}$$

$$\tan 2\Omega = \frac{a_1^2 b^2 \sin 2\alpha + a^2 b_1^2 \sin 2\beta}{a_1^2 b^2 \cos 2\alpha + a^2 b_1^2 \cos 2\beta}$$

$$= \frac{\left(\frac{a_1}{b_1}\right)^2 . \sin 2\alpha + \left(\frac{a}{b}\right)^2 . \sin 2\beta}{\left(\frac{a_1}{b_1}\right)^2 . \cos 2\alpha + \left(\frac{a}{b}\right)^2 . \sin 2\beta} \quad \ldots \ldots (8)$$

$$= \frac{m + m'}{n + n'} \text{ (for computation).}$$

To obtain the mean motion, and time of perihelion passage, two position-angles near the extreme dates are chosen from the interpolating curve, and then by means of the formula

$$\tan(\theta - \Omega) = \cos \gamma \tan(v + \lambda)$$

two corresponding values of the true anomalies are found, which yield two mean anomalies, u and u', from

$$\tan \frac{u}{2} = \sqrt{\frac{1-e}{1+e}} . \tan \frac{v}{2}$$

If we put $\quad u - e . \sin u = n(t - T) = A$

and $\quad u' - e . \sin u' = n(t' - T) = A'$

then $\quad T = \dfrac{A't - At'}{A' - A}$ and $\quad n = \dfrac{A' - A}{t' - t}$

Space will not allow of reproducing the measures made of a binary star. Hence I take for an example of the application of these formulæ the well-known binary ζ Herculis. The measures will be found on pp. 472 and 473 of volume lvi. of the 'Memoirs of the Royal Astronomical Society.' For our purpose we may commence with the measures of 1870, and referring to p. 465 of the same volume we shall see that annual means were obtained and plotted on cross-ruled paper. From the curves drawn through these annual means a series of positions for the commencement of each year were found as follows :—

Date.	Angle.	Distance.		Date.	Angle.	Distance.
	°	''			°	''
1870	195·0	1·11		1888......	79·0	1·57
71......	187·0	1·18		89......	75·0	1·53
72......	178·5	1·25		90......	71·0	1·44
73......	170·5	1·29		91......	66·5	1·38
74......	160·5	1·32		92......	60·0	1·39
75......	152·5	1·29		93......	53·0	1·39
76......	145·0	1·27		94......	46·0	1·29
77......	137·0	1·27		95......	38·5	1·09
78......	130·0	1·31		96......	27·5	0·86
79......	123·5	1·40		97......	11·0	0·65
80......	117·5	1·49		98......	328·0	0·56
81......	112·0	1·50		1899......	280·0	0·56
82......	107·0	1·49		1900......	250·5	0·65
83......	101·5	1·48		1......	233·0	0·62
84......	96·5	1·57		2......	221·0	0·96
85......	92·0	1·65		3......	210·0	1·04
86......	88·0	1·64		4......	199·5	1·10
1887......	83·5	1·58		1905......	191·5	1·16

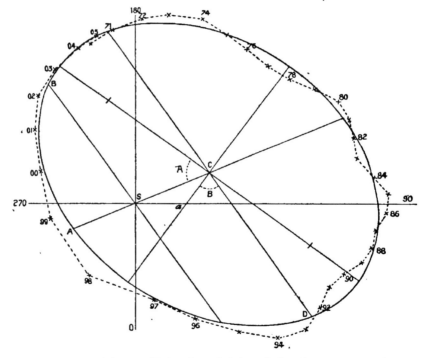

These positions will be found laid off in the accompanying diagram, and an ellipse has been drawn to represent the adopted apparent orbit. The angles should be corrected for precession, viz. :

$$0°\!\cdot\!0055 \sin a \,.\, \sec \delta \,.\, (t-t_0),$$

which, in this instance, amounts to $0°\cdot0061$ per annum, and may be neglected.

By joining up the star S with the centre we obtain the projected major axis of the true ellipse, and A the periastron.

Also, as the ratio $\dfrac{CS}{CA}$ remains constant whatever be the inclination, we at once have a value of the eccentricity, viz. :

$$e = 0\cdot556.$$

Then $\qquad \log \dfrac{1}{1-e^2} = \log \left(\dfrac{a}{b}\right)^2 = 0\cdot16254.$

If we now draw through S a chord which shall be bisected at S, this chord is the projected latus rectum, and a chord parallel through C is the projected semi-minor axis. Great care is necessary to get the required chord *. We now have

$$OSA = \alpha = 293°\cdot0 \quad \text{and} \quad OSB = \beta = 215°\cdot3$$
$$AC = a' = 0''\cdot95 \qquad CD = b' = 1''\cdot16.$$

The computation then proceeds as follows :—

L. sin 2α....	$9\cdot85693\,n$	L. sin 2β....	$9\cdot97326$
L. cos 2α....	$9\cdot84177\,n$	L. cos 2β....	$9\cdot53196$
L. $\left(\dfrac{a'}{b'}\right)^2$....	$9\cdot82504$	L. $\left(\dfrac{a}{b}\right)^2$....	$0\cdot16254$
$m = -0\cdot481$		$m' = +1\cdot367$	
$n = -0\cdot464$		$n' = +0\cdot495$	

$$
\begin{aligned}
\text{L. } (m+m') &\ldots\ldots\quad 9\cdot94743 \\
\text{L. } (n+n') &\ldots\ldots\quad 8\cdot49136 \\
\text{L. tan . } 2\,☊ &\ldots\ldots\quad 1\cdot45607 \\
☊ &= 44° \quad 0' \\
\alpha - ☊ &= 249 \quad 0 \\
\beta - ☊ &= 171 \quad 15
\end{aligned}
$$

---

* Sometimes this is troublesome. It can, however, easily be computed from the formula

$$\tan B = \frac{t^2}{c^2} . \tan A,$$

where t and c are the transverse and conjugate axes of the apparent ellipse, A the angle between t and ASC, and B the angle between c and DC, the chord required. In the present instance A is measured as $58°\cdot5$ and $\dfrac{t^2}{c^2}$ is known.

Thus $\qquad \log \dfrac{t^2}{c^2} = 0\cdot2582$

$$
\begin{aligned}
\text{L. tan.} 58°\cdot5 &= 0\cdot2127 \\
\text{L. tan B} &= 0\cdot4709 \quad \text{and} \quad B = 71° \ 12'.
\end{aligned}
$$

The angle at a is measured . . . . . $a = 54° \ 10''$.

Then $\qquad 90+B+a = $ angle of chord $= 215° \ 22'.$

	Equation (5).	*Equation* (7).
L. tan $(a - \text{☊})$	0·41582	0·41582
L. tan $(\beta - \text{☊})$ ..	9·18728 n	9·18728 n
	2)9·60310 n	2)1·22754
L. cos i	9·80155	0·61377 L. tan λ

$$i = 50^\circ\ 43' \quad \text{and} \quad \lambda = 256^\circ\ 19'$$

From equation (1)

$$\text{Log } 0'''\cdot95 = 0·97772$$
$$\text{L. sec.} \lambda = 0·62607\ n$$
$$\text{L. cos } (a - \text{☊}) = 9·55433\ n$$
$$\text{Log } a = 0·15812$$
$$\therefore \quad a = 1''\cdot44.$$

The following form (table, p. 34) for computing T and P is convenient.

In volume 18 of the 'Memoirs of the Royal Astronomical Society,' Sir John Herschel gives an analytical method of finding the orbit of a binary star. Employing, as before, the interpolating curve to obtain a number of values of θ and ϕ, for, say, every 5 degrees, he expresses them in rectangular co-ordinates by means of

$$x = \phi \cos \theta \quad \text{and} \quad y = \phi . \sin \theta.$$

Then, assuming the apparent orbit a conic section, it will be of the form

$$0 = 1 + ax + \beta y + \gamma x^2 + \delta a y + e y^2 \quad . \quad . \quad . \quad . \quad (1)$$

from which a series of equations may be obtained, and the co-efficients a, β, γ, δ, and e determined by the method of least squares, and hence the elements of the apparent ellipse. Next, by a slight adaptation of the formulæ in his previous paper, the elements of the true ellipse are deduced.

This method is the foundation of most of the analytical methods, but it is scarcely, if ever, used. Generally, after the coefficients a, β, γ, δ, and e have been determined, the custom is to pass on to the elements of the true ellipse. The method most in favour is due, according to Prof. Glasenapp, to Kowalsky, who published his formulæ in the 'Proceedings of the Kasan Imperial University,' 1873. Kowalsky takes the general equation, as used by Herschel, and, as the line of nodes lies in both the apparent and true ellipse, he turns the axis x through the angle ☊ to the line of nodes, and the axis y through i, the inclination of the orbit-plane, into the true orbit :

x becomes $x' \cos \text{☊} - y' \sin \text{☊} . \cos i$,

y ,, $x' \sin \text{☊} + y' \cos \text{☊} . \cos i.$

	1904	1875	1870
t	1904	1875	1870
θ	199°.5	152°.5	195°.0
$\tan(\theta-\Omega)$	9.6587 n	0.4755 n	9.7438 n
sec i	0.1985	0.1985	0.1985
$\tan(v+\lambda)$	9.8572 n	0.6740 n	9.9432 n
$v+\lambda$	144.35	101.97	138.80
v	−112.07	205.65	242.50
$\dfrac{v}{2}$	−56.04	102.83	121.25
$\tan\dfrac{v}{2}$	0.1715 n	0.6609 n	0.2172 n
$\sqrt{\dfrac{1-e}{1+e}}$	9.8377	9.8377	9.8377
$\tan\dfrac{u}{2}$	0.0092 n	0.4986 n	0.0549 n
u	−91.20	215.20	262.77
$e\cdot\sin u$	−31.86	−18.36	−31.60
$u-e\sin u$	A'−59.34	A 233.56	294.37
Log A' and A	1.773348 n	2.368398	2.468893
Log 1904	......	3.279667	3.279667
Log 1875	3.273001	5.648065	5.748560
Log 1870	3.271842	5.046349 n	5.045190 n
At''	......	+444698	+560480
A't	......	−111263	−110965
A't−At''	......	−555961	−671445
Log (A't−At'')	......	5.745044 n	5.827010 n
Log (A'−A)	......	2.466719 n	2.548647 n
		3.278325	3.278363
T	......	1898.2	1898.4
	Log (1904−t)	1.462398	1.531479
	Log (A'−A)	2.466919 n	2.466719 n
Log n	......	1.004321 n	1.017168 n
	Log 360	2.556303	2.556303
	......	1.551982	1.539135
P	......	35.65	34.65

These values substituted in equation (1) will give an equation which we may call (2).

Again, as the principal star is in the focus of the real orbit, we write

$$\frac{(x+ae)^2}{a^2}+\frac{y^2}{b^2}-1=0,$$

where the axis of x is the major axis.

This is inclined at an angle λ to the line of nodes, and therefore

$$\frac{(x\cos\lambda+y\sin\lambda+ae)^2}{a^2}+\frac{(-x\sin\lambda+y\cos\lambda)^2}{b^2}-1=0 \quad . \ (3)$$

By equating the coefficients of (2) and (3) the following formulæ are deduced :

$$\left.\begin{array}{l}
\dfrac{\tan^2 i}{p^2} \cdot \sin 2\,\Omega \;=\; \delta - \dfrac{\alpha\beta}{2} \\[2mm]
\dfrac{\tan^2 i}{p^4} \cdot \cos 2\,\Omega \;=\; \dfrac{\beta^2}{4} - \dfrac{\alpha^2}{4} + \gamma - \epsilon \\[2mm]
\dfrac{2}{p^2} + \dfrac{\tan^2 i}{p^2} \;=\; \dfrac{\beta^2}{4} + \dfrac{\alpha^2}{4} - (\gamma + \epsilon) \\[2mm]
e \sin \lambda \;=\; -\dfrac{p}{2}(\beta\cos\Omega - \alpha\sin\Omega).\cos i \\[2mm]
e \cos \lambda \;=\; -\dfrac{p}{2}(\beta\sin\Omega + \alpha\cos\Omega) \\[2mm]
a(1 - e^2) \;=\; p
\end{array}\right\} \quad . \quad . \quad (4)$$

As will be seen, these formulæ are easy of application when α, β, γ, δ, and ϵ are known. Prof. Glasenapp has given an excellent method * of determining these, and its first proviso falls into line with what has already been stated as the first requisite, viz. :— Draw the apparent ellipse with the utmost care. Having done this, put in equation (1), $y = 0$; then

$$\gamma x^2 + \alpha x + 1 = 0$$

the roots of which are the co-ordinates of the two points where the apparent ellipse cuts the axis of x.

If these roots be denoted by x_1 and x_2,

$$\alpha = -\frac{x_1 + x_2}{x_1 x_2} \quad \text{and} \quad \gamma = \frac{1}{x_1 x_2}.$$

Similarly, putting $x = 0$,

$$\beta = -\frac{y_1 + y_2}{y_1 y_2} \quad \text{and} \quad \epsilon = \frac{1}{y_1 y_2}.$$

The co-ordinates x_1 and x_2 and also y_1 and y_2 being of opposite sign, both γ and ϵ will be negative.

To determine δ another point $(x_3 y_3)$ must be taken such that $x_3 y_3$ has a maximum value

$$\delta = -\frac{1 + \alpha x_3 + \beta y_3 + \gamma x_3^2 + \epsilon y_3^2}{x_3 y_3}.$$

This point can usually be selected without calculation, and indeed, in practice, it will be found better to take two or three points and adopt the mean value of δ. The mean motion and time of periastron may be found from the equations already given and worked out for two dates.

* M. N. Roy. Ast. Soc., 1889 March.

Using the same apparent orbit as before, we obtain, by Glasenapp's method :—

Putting $y=0$: $\quad x_1 = +0''\cdot546$; $\quad x_2 = -1''\cdot200$

$\qquad\qquad x=0$: $\quad y_1 = +1\cdot581$; $\quad y_2 = -0\cdot512$

For δ: $\qquad\qquad x_3 = +0\cdot738$; $\quad y_3 = +1\cdot150$

$$\log a = \log -\frac{x_1+x_2}{x_1 x_2} = 9\cdot9992\,n. \qquad \log \gamma = 0\cdot1830\,n.$$

$$\log \beta = \log -\frac{y_1+y_2}{y_1 y_2} = 0\cdot1310. \qquad \log \epsilon = 0\cdot1020\,n.$$

$$\log \delta = \log -\frac{1+a x_3 + \beta y_3 + \gamma x_3^2 + \epsilon y_3^2}{x_3 y_3} = 9\cdot9070.$$

Hence

$$a = -0\cdot998; \quad a^2 = 0\cdot996; \quad \gamma = -1\cdot524; \quad \delta = +0\cdot807$$
$$\beta = +1\cdot352; \quad \beta^2 = 1\cdot828; \quad \epsilon = -1\cdot265$$

Equations (4) then become

$(a) \quad \dfrac{\tan^2 i}{p^2}\cdot\sin 2\,\Omega = +1\cdot482$ $\left.\right\}$ $\quad \log \tan 2\,\Omega = 1\cdot4633\,n$

$\qquad\qquad\qquad\qquad\qquad\qquad\qquad\qquad 2\,\Omega = 92°\ 0'$

$(b) \quad \dfrac{\tan^2 i}{p^2}\cdot\cos 2\,\Omega = -0\cdot051$ $\qquad\qquad \Omega = 46\ 0$

$(c) \qquad \dfrac{2}{p^2} + \dfrac{\tan^2 i}{p^2} = +3\cdot495$

$$\tan^2 i = +1\cdot484 \quad . \quad . \quad . \quad \text{from } (a)$$

$$\frac{2}{p^2} = +2\cdot011 \quad \text{and} \quad \log p^2 = 9\cdot9979$$

$$\tan^2 i = 1\cdot484\,p^2$$

$$i = 50°\ 52'$$

$(d) \qquad\qquad e \sin \lambda = -\dfrac{p}{2}(\beta \cos \Omega - a \sin \Omega)\cos i$

$(e) \qquad\qquad e \cos \lambda = -\dfrac{p}{2}(\beta \sin \Omega + a \cos \Omega)$

$$\tan \lambda = \frac{+0\cdot939 + 0\cdot718}{+0\cdot972 - 0\cdot693}\cdot\cos i$$

$$\lambda = 254°\ 44'$$

$$e\cdot\cos\lambda = -\frac{p}{2}\times 0\cdot279$$

$$e = 0\cdot528$$

$(f) \qquad\qquad a = \dfrac{p}{1-e^2} = 1''\cdot40$

The two sets of elements are :—

	Herschel.	Kowalsky-Glasenapp.
☊	44° 0′	46° 0′
i	50 43	50 52
λ	256 19	254 44
e	0·556	0·528
a	1″·44	1″·40

The preceding methods have been dealt with fully as being those in most general use. There are, however, various other methods, and I would recommend the perusal of the following :—

Thiele's graphical method as explained in Wilson and Gledhill's 'Handbook.' It is altogether graphical and gives good results in the hands of a competent draughtsman. I have by me some beautiful drawings made by Walter Sang in connection with an orbit of 70 Ophiuchi. Again, in *Ast. Nach.* 3336 is a graphical method, by Zwiers, founded on the use of the projection of the circle circumscribed about the true orbit. He applies his method to Sirius. H. Norris Russell, ignorant of Zwiers's paper, gives a similar method in *Popular Astronomy*, 1898 May, applying it to η Cassiopeiæ.

Another method is given by C. P. Howard in *Astronomy and Astrophysics*, 1894 June. He discusses the orbit of Sirius.

Prof. Rambaut, in Proc. Roy. Dublin Society, vol. 7, January 1891, gives a geometrical method of finding the most probable apparent orbit of a double star.

Beyond these, for the application of formulæ to irregular motion, one cannot do better than read:

Seeliger.—Fortgesetzte Untersuchungen über das mehrfache Sternsystem ζ Cancri. München, 1888.

Schorr.—Untersuchungen über die Bewegungverhältnisse in dem dreifachen Sternsystem ξ Scorpii. München, 1889.

Again, some computers, after finding a set of elements, use various means of improving them. Klinkerfues has shown an easy method of doing this, and a full description will be found in Gledhill and Wilson's 'Handbook.'

But in order to know whether the elements require improving we must have a means of comparing the values of θ and ρ given by the elements with those actually observed.

This necessitates working backwards by the formulæ

$$u - e \sin u = \mu(t - T) \quad \ldots \quad (1)$$

$$\tan \frac{u}{2} = \sqrt{\frac{1-e}{1+e}} . \tan \frac{v}{2} \quad \ldots \quad (2)$$

$$\tan(\theta - ☊) = \tan(v + \lambda) . \cos i \quad \ldots \quad (3)$$

38

To obtain u from the arguments $\mu(t-T)$ and e, I have found Åstrand's * tables most useful, where e is given up to 1·00 and u to 180°.

Let us take for example 1885·0.

Here $(t-T)$ is $-13\cdot3$ and $\mu(t-T) = +136\cdot3$

With value $e = 0\cdot556$, Åstrand gives $u = 151°\cdot5$, and we hence find $v = 164°\cdot5$ by equation (2).

Equation (3) gives

$$(\theta - \mathcal{Q}) = 48°\cdot6 \quad \text{and} \quad \therefore \; \theta = 92°\cdot0.$$

For 1875 we have

$$(t-T) = -23\cdot3 \quad \text{and} \quad u = 218°\cdot0,$$

from which we find $\theta = 145°\cdot9$.

	θ_0.	θ_c.	$c-o$.
1875	152°·5	145°·9	$-6°\cdot6$
1885	92 ·0	92 ·6	$+0$ ·6

The apparent ellipse might be readjusted to reduce the value of e. Also our period of 35·1 years is perhaps a trifle large.

The computation of ρ is from the formula

$$\rho = a(1 - e \cos u)\frac{\cos(v-\lambda)}{\cos(\theta_c - \mathcal{Q})}.$$

Personal Equation.—Certainly the persistent manner in which Otto Struve attacked this troublesome source of error entitles his results to very great consideration. Quantities determined from observations extending over a period of 20 years must be looked upon as real

His method, in brief, consisted in observing artificial double stars formed by small ivory cylinders placed in holes in a black disk, and his results are—

(1) The errors of the position-angles do not depend directly on the distances, but on the visual angles which the distances subtend, as viewed with different eyepieces.

(2) The corrections can be represented for each system of stars by a constant term, and other terms depending on the direction of the stars, with respect to the vertical.

(3) The corrections are identical for position-angles differing 180°.

From these considerations he deduced formulæ and applied the correction to his observations; and yet I would prefer his original

* 'Hülfstafeln zur leighten und genauen Auflösung des Kepler'schen Problems,' von J. J. Åstrand. Leipzig verlag vo Wilhelm Engelmann, 1890.

measures, in part because the observations of the artificial stars were made during summer afternoons, and in part because the stars were so particularly artificial.

There is much the same difference between observing an artificial and an actual star as there is between fencing with a dummy and a living enemy ; and we all know what Master Proudfute says about that :—" To tell you the truth, I strike far more freely at a helmet or bonnet when it is set on my wooden Soldan—then I am sure to fetch it down. But when there is a plume of feathers in it that nod, and two eyes gleaming fiercely from under the shadow of the visor, and *when the whole is dancing about here and there*, I acknowledge it puts my hand out of fence."

If we consider the annexed figure as a double-star system, the required measures are the angle ACE and the distance CE.

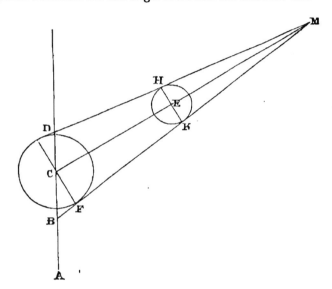

But by a bias in locating the centres C and E there is, we may consider, a tendency to measure ABK or ADH instead of ACE.

Taking the extreme case where the actual tangent is measured :

Let p represent the true position-angle ACE, then CMB or θ is the angle due to P.E. ; and if ρ is the distance CE,

$$\sin \theta = \frac{CF}{\rho + EK \cdot \mathrm{cosec}\, \theta},$$

or

$$\sin \theta = \frac{1}{\rho}\,(CF - EK),$$

which means that θ is dependent on the difference of magnitude directly, and inversely on the distance.

This distance is, of course, the apparent separation as viewed with different eyepieces, and corresponds to Otto Struve's first deduction.

We might go a step further and introduce the effect of the inclination to the vertical; but it is of doubtful utility to do so, in consequence of the observer involuntarily falling into the habit of accommodating his head to the line joining the stars, and so in effect observing at a nearly constant angle one side or other of the vertical. I should be inclined to consider the amount constant, but differing in sign from 0° to 180° to that from 180° to 360°. What happens in the immediate neighbourhood of the vertical it is difficult to say.

Considering the P.E. in measuring distances as arising from a similar bias in judging the centres, it would be some function of diameters of disks, i. e. of the magnitudes.

Personal equation, being thus dependent on magnitude, visual angles, and on the time, is different in different systems, and may even change in the same system. We are on much safer ground if we abandon formulæ and consider the smooth curve through the annual means of all observers as the most probable, and then deduce a table of differences for each observer arranged in order of time. Such a table furnishes not only a mean P.E. for an individual, but would indicate at once any change in that P.E.

I have not been able to find any experiments made by Burnham for deducing a P.E., and Hall's opinion is that "The formulæ and corrections for personal errors of observation which have been deduced and applied by some astronomers seem to me of doubtful utility, and a better, I think, is to compare the measurements of the same stars by different observers."

The above was written in 1893, and I see no reason to modify it. From the very nature of the measures it is evident personality exists, but in an ever varying amount, not only as between two observers but between the same observer on different occasions, and even in his measures of the same system. It is more or less to be considered as an accidental error. I had not intended writing further on the subject, but the recent experiments of MM. Salet and Bosler at the Paris Observatory seem so practical, and their remarks so pertinent, that I venture to quote from them. It would seem that the irregularities in the systems of ζ Cancri and 70 Ophiuchi drew their attention to the subject. Employing the equatorial of the *tour de l'Est* (aperture 0·38 m. and 8 m. focus), they first established the existence of a P.E. between themselves and also one due to position. They then eliminated the P.E. by means of a reversible prism. They remark:—

"L'équation personnelle, en effet, *n'est pas constante*. On conçoit qu'il n'y a aucune raison pour qu'elle le soit et de plus nous l'avons vérifié expérimentalement par a travail. Les différences entre les lectures obtenues avec et sans prisme sont en effet très variables suivant les jours d'observations. L'équation personnelle dépend, comme on l'a déjà remarqué, pour un même couple, de l'éclat et de la définition des images ; elle dépend aussi de l'heure de l'observation, heure variable avec la saison pour un même angle horaire, et qui fait changer l'éclat relatif des étoiles et du fond du ciel. Enfin l'équation personnelle dépend surtout de la position et de la fatigue de l'observateur " *.

There are several other points in connection with double-stars in addition to those already mentioned, such as :—Light ratio, magnitude, masses, orbit-planes, &c.

The intensity of the light of stars varies, and an arbitrary scale is adopted to express their relative light. The scale in general use is known as Pogson's Scale, where the ratio of light between a star of magnitude $(m-1)$ and one of magnitude m is 2·512. This 2·512, of which the logarithm is ·4, is known as the " Light Ratio."

Put generally the formula becomes

$$L_m = (2·512)^d L_{m+d} \quad \text{or} \quad \log \frac{L_m}{L_{m+d}} = 0·4d.$$

Suppose, for instance, we wish to know the difference between the light of the two components of ζ Herculis whose magnitudes are 3 and 6 :

$$L_3 = (2·512)^3 L_6 = 16 L_6,$$

or the principal star gives 16 times the light of the companion.

This formula becomes of service in cases where we know the magnitudes of each and wish to know the combined magnitude ; or if we know the combined magnitude and the magnitude of either we can deduce the magnitude of the other.

In the above example the combined light is to the light of the principal star as 17 : 16.

But $\quad L_C = (2·512)^{A-C} L_A \quad$ or $\quad \dfrac{17}{16} = (2·512)^{A-C}$

$\qquad A - C = 0·065$

but $\quad A = 3·0 \qquad \therefore \ C = 2·93.$

Generally, if we know any two of the quantities, A, B–A, C, we can find the third. The following table will give rough values :—

B–A.	A–C.	B–A.	A–C.
0·1	0·7	1·2	0·3
0·4	0·6	1·8	0·2
0·6	0·5	3·0	0·1
0·9	0·4	4·0	0·0

Suppose we know that the combined light of the two components of κ Pegasi is 3·8, and that the principal star is 4·3. We have

$$A - C = 0·5$$

and therefore from the table

$$B - A = 0·6 \quad \text{or} \quad B = 4·4.$$

* *Bulletin Astronomique*, Janvier 1908, pp. 18–26.

The formulæ connecting the masses, parallaxes, semi-axis major of binary stars are obtained as follows :—

Let P = period in years.

  p = parallax in arc.

  a = semi-major axis in arc.

  d = distance between components in ast. units.

  M = mass of Sun, the Earth's mass being neglected.

m and m' = the masses of the components.

Then

$$P^2 : (\text{Earth's period})^2 : : \frac{d^3}{m+m'} : \frac{(\text{Earth's distance})^3}{M}$$

$$m+m' = \frac{d^3}{p^2} . M \quad \text{and} \quad d = \frac{a}{p}$$

$$\therefore \quad m+m' = \frac{a^3}{p^3 . P^2} . M.$$

Thus from the table on p. 44, for Sirius,

$$m+m' = \frac{8^3}{(0\cdot38)^3 . 52^2} . \text{ times the Sun's mass}$$

$$= 3\cdot47 \text{ times the Sun's mass}$$

$$m = 2\cdot3 \text{ mass of Sun} \quad \text{and} \quad m' = 1\cdot2 \text{ mass of Sun}.$$

Again, as the light from a star of parallax $1''\cdot0$ takes $3\cdot26$ years to reach us, it follows that the distance of any star from our system, in light-years, is

$$\frac{3\cdot26}{\text{parallax}}.$$

Sirius is distant $8\cdot6$ light-years.

Further, the motion across the line of sight (P.M.) may be converted into miles per second from

$$\text{Motion across line of sight} = \frac{9}{10}\{\text{P.M.} \times \text{light-years}\}$$

or $\text{motion across line of sight} = 2\cdot94 \dfrac{\text{proper motion}}{\text{parallax}}.$

Masses of Binary Stars.—From the micrometer measures alone we cannot obtain any clue as to the masses of the stars composing a binary system ; but where we can secure observations of one star independently of the other we have a means of deducing their relative masses. Thus, the micrometer measures give us the orbit of one star relative to the other. Meridian observations of one star will give its orbit about the common centre of gravity. Having these two we can find the real orbit of the second star.

A comparison of the two orbits about the centre of gravity gives the relative masses.

From the nature of the case we cannot hope to apply this to many pairs. The table on p. 44 is taken from the Introduction to Lewis's Memoir *.

It is evident, from this table, that magnitude is not the criterion of mass ; the relative colours and relative masses appear much more in- dependent.

This statement naturally brings to mind two relations between the colours and magnitudes of binary stars which for many years have been accepted, viz. :—

(1) When the magnitudes of the components differ but little, their colours differ little.
(2) When there is a considerable range in the magnitudes, the difference in colour is very marked.

Some years ago Mr. Gore, from a discussion of 60 binaries, found these statements held true in general.

In the *Observatory*, 1906 August, Mr. Lewis extends the rules to pairs physically connected as follows :—

Physical pairs in the ' Mensuræ Micrometricæ.'

No. of pairs.		Colour of Principal.	Colour of Companion.	Mean difference of magnitude. m.
20	....	Y W	Y W	0·5
185	....	W	W	0·6
63	....	Y	Y	0·7
10	....	W	Y	0·8
14	....	W	B W	1·1
11	....	Y	W	1·9
60	....	W	B	2·2
21	....	Y W	B	2·4
106	....	Y	B	2·5

These are mostly in the Northern Hemisphere. From 50 stars in the Southern Hemisphere and not in Struve, Mr. Lewis obtained

Colour of Principal.	Colour of Companion.	Mean difference of magnitude. m.
W	W	0·5
Y	Y	0·6
W	B	2·9
Y	B	2·6

a sufficient agreement with the above.

* Memoirs R. A. S. vol. lvi.

	Period in years	Parallax	Semi-axis major		Distance, light-years	Proper motion	Magnitudes	Light Ratio	Relative masses	Colours	Authority for masses
			Arc.	Ast. units.							
η Cassiopeiae	233	″.28	″8.5	30.4	11.6	″1.20	4.0 / 7.6	26 to 1	2 to 1	Y P	Lewis.
ε Hydrae	15	...	0.3	...	...	0.20	3.0 / 6.0	16 ,, 1	1 ,, 6	Y B	Lewis.
σ Coronae	340	...	3.8	...	23.3	0.34	5.0 / 6.0	2.5 ,, 1	1 ,, 4	Y B	Lewis.
ζ Herculis	34	.14	1.4	10.0	...	0.61	3.0 / 6.0	16 ,, 1	1 ,, 1	Y B	Lewis.
λ Ophiuchi	134	...	1.0	...	...	0.31	4.0 / 6.1	6.5 ,, 1	3 ,, 13	Y B	Lewis.
70 Ophiuchi	88	.16	4.6	31.0	20.4	1.12	4.5 / 6.0	4 ,, 1	1 ,, 4	Y P	Prey.
ξ Boötis	137	...	5.3	18.0	...	0.16	4.5 / 6.5	6.3 ,, 1	4 ,, 5	Y P	Bowyer.
85 Pegasi	25	.05	0.9	...	65.2	1.28	6.0 / 10.0	40 ,, 1	1 ,, 4	Y B	Furner.
,,	...	...	...	...	...	...	,, / ,,	40 ,, 1	1 ,, 3	Y B	Lewis.
Procyon	40	.27	5.0	18.5	10.9	1.22	1.0 / 12.0	24000 ,, 1	5 ,, 1	Y G	See.
40 o² Eridani	180	.22	6.2	28.4	14.8	4.05	9.2 / 10.2	2.5 ,, 1	12 ,, 13	B B	Lewis.
25 Can. Ven.	220	...	1.1	21.1	...	0.11	5.0 / 8.5	25 ,, 1	1 ,, 2	W B	Furner.
Sirius	52	.38	8.0	...	8.6	1.30	1.0 / 10.0	3980 ,, 1	2 ,, 1	W Y	See.
ξ Scorpii	44	...	0.7	...	...	0.12	5.0 / 5.2	1.2 ,, 1	3 ,, 4	Y Y	Schorr.
ζ Cancri	60	...	0.9	...	...	0.12	5.5 / 6.2	2 ,, 1	1 ,, 1	Y Y	Seeliger.
ξ Ursae Maj.	60	...	2.5	...	...	0.74	4.0 / 5.0	2.5 ,, 1	2 ,, 3	Y Y	Bowyer.
γ Virginis	182	.05	3.9	78.0	65.2	0.58	3.0 / 3.2	1.2 ,, 1	1 ,, 1	Y Y	Lewis.
α Centauri	81	.76	17.7	23.3	4.3	3.69	1.0 / 2.0	2.5 ,, 1	19 ,, 20	Y Y	Gill.

Poles of Double-star Orbits.—At various times investigations have been made with the object of finding whether the poles of double-star orbits showed any marked peculiarity of distribution. Would there, for instance, be a tendency for the axes of revolution to be parallel to the Galaxy? Such investigations, by Dr. Doberck, will be found in the *Astronomische Nachrichten*; by Dr. See in 'Stellar Systems'; and by Miss Everett in *M. N.* lvi. Nothing of a decided character could be gathered by these investigators.

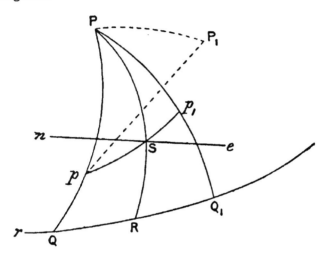

The inclination of the orbital plane to the plane of projection can be found; but this angle (i) is also the inclination of a plane inclined ($2i$) to the plane of the orbit. We thus have two sets of axes and poles and the inability to distinguish between the real and the spurious set is the great difficulty attending this research. Quite recently * Prof. Turner and Mr. Lewis, using the real and spurious poles of 59 orbits, found evidence slightly in favour of a grouping of poles near the Milky Way; but Karl Bohlin, by making choice of poles, comes to the opposite conclusion †.

The formulæ for computation are as follows (see fig.):—

Let S be the star, p and p_1 projected poles of orbit.

Then since S is also the projection of the pole of the projection plane, pS or p_1S is the angle between the poles or planes.

Let nSe be the line of nodes.

Then PSn = ☊, if P be the pole of the heavens.

* Monthly Notices, 1907 June.
† Archiv für Matematik, Astronomie och Fysik, Stockholm, Band 3, No. 19.

Let a and δ be the R.A. and Dec. of star.

A and D   „   „   pole p.

A_1 and D_1 „   „   pole p_1.

$$\cos(90 + \Omega) = \frac{\cos pP - \cos PS \cdot \cos pS}{\sin PS \cdot \sin pS'}$$

$$\sin D = \cos i \cdot \sin \delta - \sin i \cdot \cos \delta \cdot \sin \Omega$$

$$\sin D_1 = \cos i \cdot \sin \delta + \sin i \cdot \cos \delta \cdot \sin \Omega$$

also
$$\sin(a - A) = \sin i \cdot \cos \Omega \cdot \sec D$$

$$\sin(a - A_1) = -\sin i \cdot \cos \Omega \cdot \sec D_1.$$

Again, if P_1 be the pole of the Galaxy *, pP_1 is the inclination of plane of orbit to plane of Galaxy.

Let a_1 and δ_1 be the R.A. and Dec. of the Galactic Pole,

$$\cos(a_1 - A) = \frac{\cos I - \sin D \cdot \sin \delta}{\cos D \cdot \cos \delta_1}$$

$$\cos I = \sin D \cdot \sin \delta_1 + \cos D \cdot \cos \delta_1 \cos(a_1 - A).$$

T. Lewis.

---

* Sir J. Herschel's positions of the poles of the Milky Way are:—

	R.A.	Dec.
North Pole	$12^h\ 47^m$	$+27°$
South Pole	$0\quad 47$	-27

Printed by Taylor & Francis, Red Lion Court, Fleet Street, E.C.

Breinigsville, PA USA
12 September 2010
245230BV00003B/4/P

9 781146 280266